Clouds

by Margie Burton, Cathy French, and Tammy Jones

There are many kinds of clouds.
They are not all the same.

Take a look up in the sky.
Can you see some clouds?
Do you like to look at them?

The clouds are made from drops of water. The drops of water are too little for us to see.
When the air gets cool, the drops of water make clouds.

Some clouds are up high in the sky. Other clouds are down low. How many kinds of clouds do you see?

These clouds look like feathers.

These clouds look like cotton.

These clouds look like thick, gray blankets.

Look at these clouds.

These clouds do not make rain.
It is a good day to play outside.

These clouds are up very high in the sky.

Look at these clouds.
They look like cotton.
Can you see some shapes
in these clouds?

These clouds are up very high in the sky.

Sometimes these same clouds look very dark.

What do you think is coming?

When these clouds look dark, they tell us that rain is coming.

Look at these clouds.

These clouds look gray and flat. It is a rainy day. Rain or snow may fall for a long time.

rain

snow

We can find clouds up high in the sky, in the middle of the sky, and down low in the sky. Let's look and see.

Where are the clouds in the sky?	What do they look like?	What are they called?
high in the sky	feathers	cirrus
middle of the sky	cotton	cumulus
low in the sky	thick, gray blankets	stratus

Let's make a cloud.

1. Put some very warm water into a jar.

2. Place a metal lid upside down on top of the jar.

3. Put some ice cubes onto the lid.

What do you see? You made a cloud!